HACKING THE SAT MATH SECTION

BY: ASHWIN RAM

PROVEN TECHNIQUES TO GET A HIGH SCORE

@Teach4Success.com

TABLE OF CONTENTS

                    @Teach4Success.com

CHAPTER 1: NAILING EXPONENTS AND RADICALS

The first and foremost thing to know on the SAT Math Section is how to nail exponents and radicals. Make sure to know the basic laws as follows:

Law	Example
x^1 = x	3^1 = 3
x^0 = 1	3^0 = 1
x^m * x^n = x^(m+n)	3^2 * 3^5 = 3^(7)
x^m / (x^n)= x^(m-n)	3^7 / (3^2) = (3^5)
(x^m)^n = x^(mn)	(3^6)^2 = 3^12
(xy)^m = x^m * y^m	(6)^5 = (2 * 3)^5 = 2^5 * 3^5
(x/y)^m = x^m / (y^m)	(⅔)^4 = 2^4 / (3^4)
x^(-m) = 1/(x^m)	3^-4 = 1/ (3^4)

In addition, know PEMDAS and rules of precedence. Know that (-3)^2 = 9 and -3^2 = -9. Whenever you have an even exponent, this will always be true. However, with an odd exponent, you get the following:

(-3)^3 = -27 and -3^3 = -27.

Now that you all are familiar with the exponent's rules, let us try a few practice problems:

   (1)   If 3^(x+2) = y, then what is the value of 3^x in terms of y?
(a) y + 9 (b) y-9 (c) y/3 (d) y/9

In this case, the correct answer is D. 3^(x+2) = 3^(x)* 3^(2) = 3^x * 9 = y. **Thus, 3^x = y/9**

   (2)   If 3^(a+1) = 3^(-a + 7), what is a?

In this case, the correct answer is 3. Since the bases of the exponents are both 3, we can set the exponents equal to one another. Thus,

```
a + 1 = -a + 7

2a = 6
```

a = 3

    (3)   If 2a-b = 4, what is 4^a / 2^b?

The correct answer is 16. The reason is because 4 can be rewritten as 2 squared. Thus, 2^2a / 2^b. Using the law of exponential division, 2^2a / 2^b becomes 2^(2a-b). Knowing that 2a-b = 4 gets us 2^4, which equals **16.**

Now, remember that square roots are just fractional exponents. The square root of x is x^($\frac{1}{2}$). $\sqrt{(x^5)}$ is x^(5/2).

Now that we have covered all the basic laws, let us now try some official SAT problems.

1

If $a^{-\frac{1}{2}} = 3$, what is the value of a ?

  A) -9

  B) $\dfrac{1}{9}$

  C) $\dfrac{1}{3}$

  D) 9

The best answer to this problem is **B.** To eliminate an exponent from a variable simply exponentiate both sides by the reciprocal of the exponent. This gets,

```
A = 3^-2
```

A = 1/9

Let us now try another problem:

6

If $3^x = 10$, what is the value of 3^{x-3} ?

A) $\dfrac{10}{3}$

B) $\dfrac{10}{9}$

C) $\dfrac{10}{27}$

D) $\dfrac{27}{10}$

```
The best answer to this problem is C. This is because 3^(x-3) is 3^x / 3^3 by
the laws of exponents. This gets 3^x / 27. From the problem, if 3^x = 10,
this equation therefore gets 10/27 by substitution.
```

```
The next problem is very difficult. But, we can solve it!
```

13

If $2^{x+3} - 2^x = k(2^x)$, what is the value of k ?

A) 3

B) 5

C) 7

D) 8

```
The best answer to this problem is C. This is because rewriting 2^(x+3) gets
2^x * 2^3 = 2^x * 8. This gets 8(2^x) - (2^x) = k(2^x). We can therefore
factor out a (2^x) from all terms. This gets us 8 - 1 = k. Thus, 7 = k.
```

                    @Teach4Success.com

CHAPTER 2: NAILING EQUATIONS

It is a necessity to nail your equations in order to score high on the SAT.

For example,

⅓ x + 11 = 20. Rearranging gets ⅓ x = 9, so x = 27.

Systems of Equations

The SAT LOVES Systems of Equations. Currently, to make this test much more difficult, the SAT has begun incorporating Systems into their exams. Be sure to know how to solve systems like the following:

$$2x - y = 2y - 5$$
$$x + 5y = 3$$

If we manipulate the second equation, we get $x = 3 - 5y$. Rearranging the first equation gets $2x - 3y = -5$. Substituting $x = 3 - 5y$ into $2x - 3y = -5$ gets
$2(3 - 5y) - 3y = -5$. This gets $6 - 10y - 3y = -5$. This gets $6 - 13y = -5$. This gets $-13y = -11$. Thus, $y = 11/13$. Plugging this into $x = 3 - 5y$, gets
$3 - 5 (11/13)$, which comes out to $-16/13$.

You should also recognize when a system has no solutions. This entails that the slopes of both lines are the same because the lines are parallel and do not intersect. **If the coefficients on x and y are the same and they have a different y-intercept, then you are dealing with parallel lines with no solution. If the coefficients on x and y are the same and they have the same y-intercept, then you are dealing with the same line with infinite solutions.**

For example, take a look at the following problem:

¼ x - 1/12 y = 7
12x - 4y = 10.

If we multiply the first equation by 48, we get:

12x - 4y = 336. Since the coefficients of both equations 12x - 4y = 336 and 12x - 4y = 10 are the same, yet the numbers they are equivalent to are different, the system has no solutions.

                                      @Teach4Success.com

Manipulate Equations

Be able to rearrange equations to save yourself time. For example:

If ⅛ x + ¼ y = 10, what is 2x + 4y?

In this given problem, if we multiply **⅛ x + ¼ y = 10** by 16, we get
2x + 4y = **160**.

Factoring Quadratic Equations

It is also a necessity to be able to factor quadratics on the SAT. A
quadratic is an expression or equation that has a single squared term as its
highest power. For example:

$x^2 - 3x = 70$. In this equation, we must move 70 over to the other side. This
gets $x^2 - 3x - 70 = 0$.

This gets: $(x-10)(x+7) = 0$,

Which makes: $x = 10$, and $x = -7$.

Also, be able to recognize perfect square quadratics:

$(x^2 + 6x + 9) = (x+3)^2$

and

$(x^2 - 8x + 16) = (x-4)^2$

Also, be able to recognize difference of squares:

$x^2 - y^2 = (x+y)(x-y)$.

This also works with any even exponents:

$x^4 - y^4 = (x^2 + y^2)(x^2 - y^2)$

$= (x^2 + y^2)(x+y)(x-y)$

Know the quadratic formula and its applications.

For example:

$$x^2 - 3x = 70 \Leftrightarrow x^2 - 3x - 70 = 0$$

$$x = \frac{-b \pm \sqrt{b^2 - 4ac}}{2a}$$

$$x = \frac{3 \pm \sqrt{-3^2 - 4(1)(-70)}}{2(1)}$$

$$x = \frac{3 \pm \sqrt{289}}{2(1)}$$

$$x = \frac{3 \pm 17}{2(1)}$$

$$x = 10 \ \ or \ -7$$

Note, many times on the SAT They will ask for the sum or product of the two solutions. Those have basic formulas:

Sum: -b/a. In this example, -(-3)/1 = 3

Product: c/a. In this example, -70/1 = -70

Discriminant and Its Implications

The discriminant of a given quadratic equation is given by **b^2 - 4ac.**

NOTE THAT ROOTS ARE ZEROS WHICH ARE SOLUTIONS. ALL ARE SYNONYMS

                    @Teach4Success.com

Type of Discriminant	Meaning
Positive Discriminant	2 real solutions (zeros)
Discriminant = 0	1 real solution (zero)
Negative Discriminant	No real solutions

For example: x^2 - 8x - 8 = 0.

Using our discriminant formula,

b^2 - 4ac

We can see that (-8)^2 -4(1)(-8) = 64 + 32 = 96. Because 96 is positive, there are 2 real solutions!

Absolute Value Equations

Note that absolute value Equations can never equal 0.

So,

|x+2| + 4 ≠ 0

Translate Word Problems

You must be able to turn equations into word problems and vice versa. The cost of a banana is $1 and an apple is $2. You purchased a total of $100 in apples and bananas. The total number of apples and bananas is 50. Setup a system of equations.

From this,

```
a + b = 50      This is the total number equation
b + 2a = 100    Cost equations
```

Thus, subtracting the two equations, -a = -50. So, a = 50, and thus b = 0.

Now let us try problems:

                     @Teach4Success.com

1

If $a^2 + 15a = 76$ and $a > 0$, what is the value of $a + 5$?

```
Moving 76 over, a^2 + 15a - 76 = 0. Then factoring:
```

```
Thus, (a-4)(a+19)= 0, so a = 4 and -19. a > 0, so a = 4. 4 + 5 = 9.
Therefore, a = 9.
```

2

Based on the system of equations below, what is the value of $x + y$?

$$\begin{cases} 4x - 5 - 3y = 2x + 4 \\ x + 5y + 1 = y \end{cases}$$

(A) -4

(B) -2

(C) 2

(D) 4

```
Manipulating the first equation gets us: 2x - 3y = 9. Manipulating the second
equation gets us x+4y = -1.
```

```
Thus:
2x - 3y = 9
X + 4y = -1
```

```
Multiplying the second above equation by 2 gets.
```

```
2x-3y = 9
2x + 8y = -2.
```

```
Subtracting the two equations gets:
-11 = 11, so thus y = -1. Plugging this in, we get x = 3.
```

   @Teach4Success.com

Chapter 3: Acing Expressions

When you have certain fractions, be sure to employ what is known as a "fancy one." For instance, to get a common denominator of 2x, you may multiply by 2x/2x.

This is shown as follows:

$$\frac{5x}{x+2} - \frac{10}{x-2}$$

$$\frac{5x(x-2)}{x+2(x-2)} - \frac{10(x+2)}{x-2(x+2)}$$

$$\frac{5x(x-2)-10(x+2)}{(x+2)(x-2)}$$

$$\frac{\left(5x^2-10x\right)-\left(10x+20\right)}{(x+2)(x-2)}$$

$$\frac{5x^2-20x+20}{(x+2)(x-2)}$$

Notice how the common denominator was (x+2)(x-2). Multiplying each one by that eliminated a certain expression.

<u>You must also know operations such as:</u>

2x-3(4x+5).

This gets by the distributive property,

2x-12x-15 = **-10x - 15**.

<u>Be able to separate fractions:</u>
For example:
(y+1)/x can be written as y/x + 1/x.

<u>Know how to rationalize expressions as well</u>
Rationalization is the same thing as multiplying by a "fancy 1" except you multiply to eliminate a square root from the denominator.

For example:

$$\frac{2x+3x}{x-\sqrt{2}}$$

$$\frac{2x+3x}{x-\sqrt{2}}\left(\frac{x+\sqrt{2}}{x+\sqrt{2}}\right)$$

$$\frac{2x^2+2x\sqrt{2}+3x^2+3x\sqrt{2}}{x^2+2}$$

$$\frac{5x^2+5x\sqrt{2}}{x^2+2}$$

In this example, we multiply both the top and bottom by a radical expression.
Know how to cross multiply

Given the problem:

$$\frac{3}{x-5}=\frac{7}{x+3}$$

Cross multiplying gets us:

3(x+3) = 7(x-5) =

3x+9 = 7x-35   =

4x = 44   =

X = 11

Know how to substitute expressions:

If x = 5a + b and y = 10b + a what is x + y?

The correct answer is 6a + 11b

Know Expression addition as well:

$$\left(x^3y^2+6y^3-10x^2y^3\right)-\left(2x^3y^2+6x^2y^3+6y^3\right)$$

$$\left(x^3y^2+-10x^2y^3\right)+\left(-2x^3y^2+-6x^2y^3\right)$$

$$-x^3y^2+-16x^2y^3$$

Also remember polynomial multiplication:

```
(2x^2 + x + y)(x+3y) = 2x^3 +6(x^2)y + x^2 + 3xy + xy + 3y^2. Which gets:
```

2x^3 +6(x^2)y + x^2 + 4xy + 3y^2

<u>Also know how to divide polynomials that may seem very weird.</u>

$$x^4 + 5x^3 + bxy + 3y$$

If the polynomial is divisible by $x + 5$, what is the value of b (where b is a constant)?

```
In this example, we must factor out x^3 from the first expression, giving us:

x^3 (x + 5) + bxy + 3y. This gives us:

x^3(x+5) + y(bx+3).

In order for this to be divisible by (x+5), y(bx+3) must be divisible by
(x+5). Thus,

y(bx+3) = by(x+5) =

bxy + 3y = bxy + 5by =

3y -5by = 0

3y(1- 5/3 b) = 0.

Thus,

5/3 b = 1, so b = ⅗.
```

```
Now that we have covered everything, let us try some official SAT practice.
```

1

If the expression $\dfrac{5-9x^2}{2-3x}$ is written in the equivalent form $\dfrac{1}{2-3x}+A$, what is A in terms of x ?

(A) $2-3x$

(B) $2+3x$

(C) $9x^2$

(D) 5

The correct answer is B. Here is a very simple strategy to solve this. Plug in 0 to the first expression. This gives you 5/2. Plugging 0 in for x in the second expression, you get ½ + A. Therefore, ½ + A must equal 5/2. Therefore, A = 2. Essentially, A = 2, when x = 0. Therefore, we can eliminate any answer choices that do not equal 2 when x = 0. When plugging in 0, only answer choices (A) and (B) get you 2.

So now you are down to A and B. Now, we must plug in 1. This gets 5-9/ (2-3), which gets -4/-1 = 4. Plugging 1 into the next equation, we get 1/(2-3) + A = -1 + A. Therefore -1 + A = 4. So, A = 5. **Essentially, when x = 1, A must be 5.**

When plugging in 1 to answer choices A and B, only B gets you 5. Thus, B is correct.

Chapter 4: Nailing Inequalities

Multiplying by a negative

When you multiply or divide an inequality by a negative number, the sign of the inequality flips. For instance, look at the inequality below and watch what happens as we multiply by -2.

-3 < x + 3 < 3

6 > -2(x+3) > -6

This gets -6< -2x-6 < 6, getting 0 < -2x < 12. Dividing everything by -2 to get x by itself gives,

0 > x > -6 or **-6 < x < 0**

Be able to read word problems and convert to inequalities:

A banana has 98 calories. An apple has 77 calories. The daily recommended calories from fruits is 400 calories. Write an inequality that represents the possible number of bananas (b) and number of apples (a) to meet or exceed the recommended daily fruit calorie intake

98b + 77a ≥ 400.

Be able to graph inequalities:

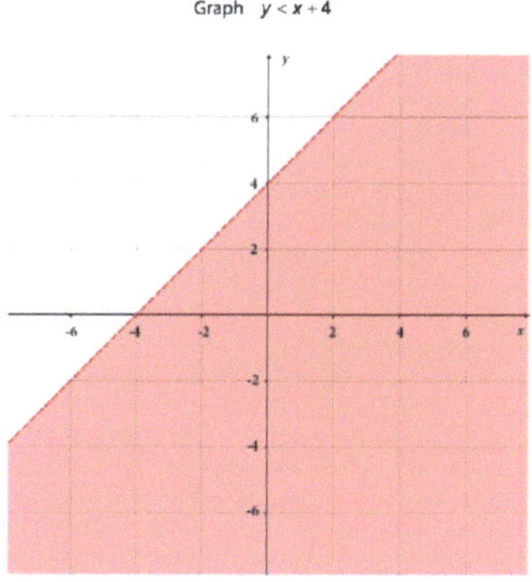

Now, let us try some practice problems!

                                    @Teach4Success.com

1

John wants to buy at least 50 bottles of soda for a birthday party. Which of the following inequalities represents the possible number of six-packs, s, and four-packs f, of soda John should buy?

(A) $6s + 4f > 50$

(B) $6s + 4f \geq 50$

(C) $\dfrac{6}{s} + \dfrac{4}{f} > 50$

(D) $\dfrac{6}{s} + \dfrac{4}{f} \geq 50$

```
This is an easy problem. For every  1 six-pack you get six sodas. Thus the
answer is B. Moreover, you can eliminate A and C immediately because the
problem asks for at least which entails greater than or equal to.
```

2

If $-\dfrac{3}{4} < 1 - 2t < -\dfrac{1}{4}$, what is one possible value of $8t - 4$?

```
To get 8t - 4, we must multiply (1-2t) by -4. However, we must multiply the
entire thing by -4. This would get the signs flipped. Therefore,
```

```
1 < 8t - 4 < 3. Therefore, one possible value would be 2. However, any answer
between 1 and 3 would work.
```


Chapter 5: Nailing Data Analysis

On the SAT, this covers a significant portion of the exam. First, let us take a look at an example. Look at the table below:

College Major	Gender		Total
	Men	Women	
Humanities	4	10	14
Natural Sciences	11	10	21
Social Sciences	8	14	22
Total	23	34	57

Given this complex table, we could be asked a multitude of questions. For example,

(1)   **What fraction of humanities and social sciences majors are female?**

In this question, remember to only look at the total amount of THESE TWO majors combined. Thus, you would get 36 total. In terms of females, adding 10 and 14 get 24. This gets you 24/36 = ⅔.

(2)   **Based on the national average that 50% of humanities majors go on to do graduate school, how many women can we expect to do graduate school from the humanities majors listed in the table (round to the nearest whole number)?**

For this problem, all we have to do is take 50% of the female humanities major. 50% of 10 is 5. Thus, 5 would go on to do this in college.

Other things to note:

- Participants in a survey/experiment must be randomly selected in order to generalize results to the greater population.
  - Ex. If only people who visit the hospital are surveyed, the results cannot be generalized to the population of an entire town.
- Participants in a survey/experiment must be randomly assigned to treatment groups in order to draw conclusions about cause and effect.
  - Ex. If people with a blood sugar above 126mg/dL are given a certain diabetes medication, but people with a blood sugar below 126mg/dL are not given that diabetes medication, a conclusion about the effectiveness of the treatment in the entire population

cannot be drawn. Instead, researchers should randomly assign the treatment without any bias in order to make conclusions about cause and effect.

Practice:

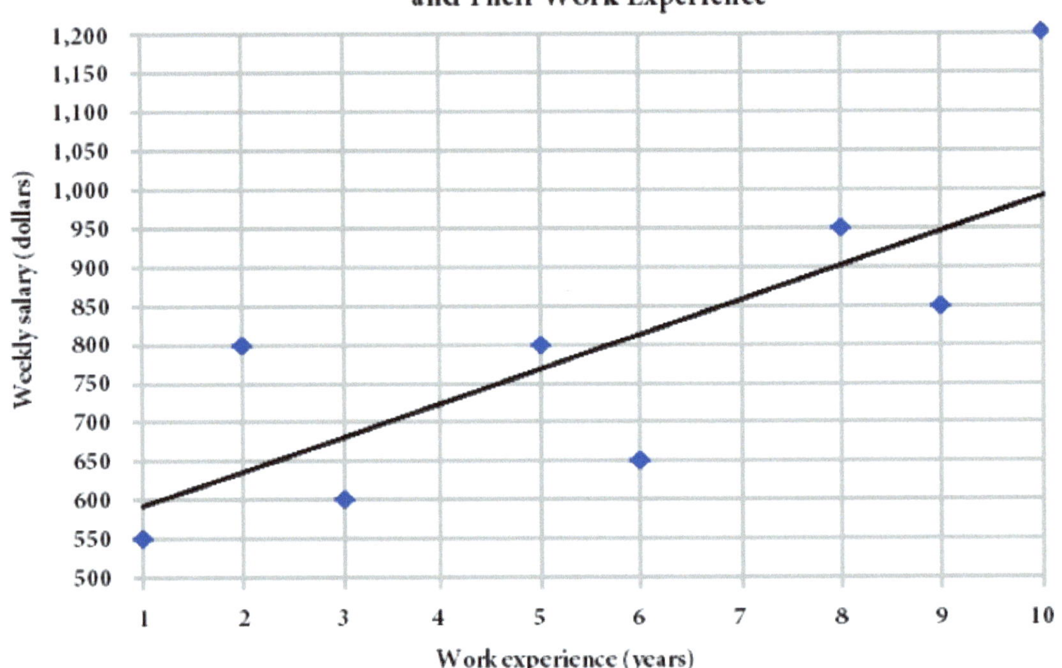

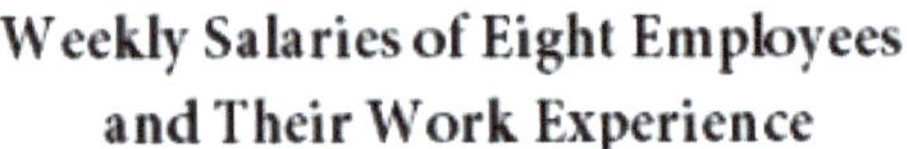

1

How many of the eight employees have an actual weekly salary that differs by more than $150 from the weekly salary predicted by the line of best fit?

(A) 2

(B) 3

(C) 4

(D) 6

The correct answer to this question is B. This is because the work experience for 2 years, 6 years, and 10 years, all differ by more than $150 from the best fit line.

Another:

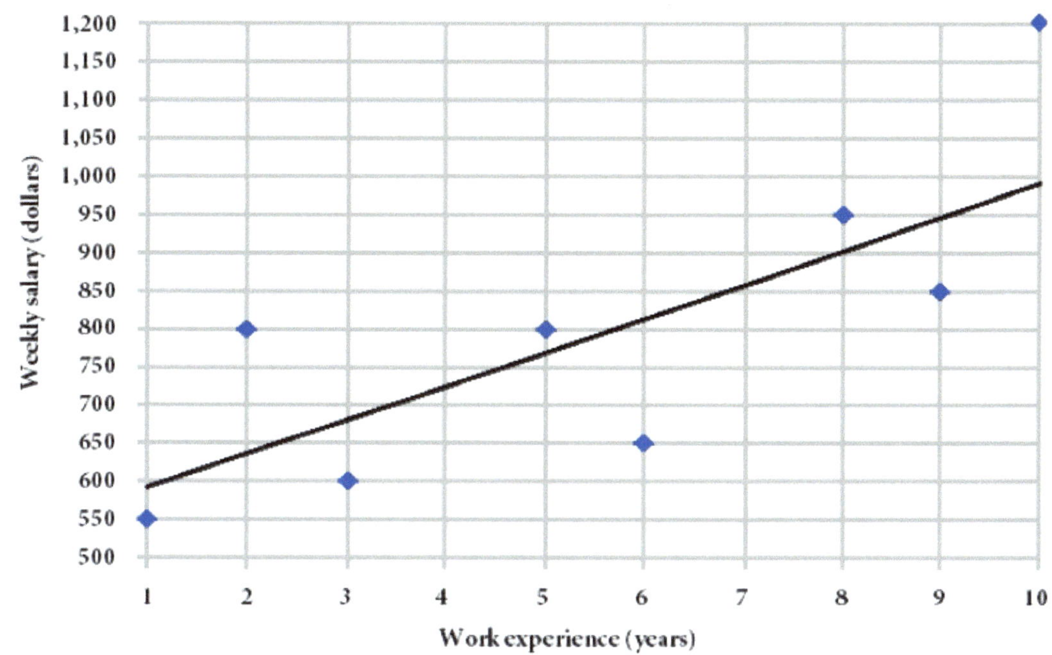

2

Which of the following is the best interpretation of the slope of the line of best fit in the context of this problem?

(A) The predicted weekly salary in dollars of an employee with 0 years of work experience.

(B) The predicted work experience in years of an employee with $0 weekly salary.

(C) The predicted weekly salary increase in dollars for 1 year increase in work experience.

(D) The work experience increase in years for every dollar increase in weekly salary.

The slope of a graph is defined by the increase in y for every increase in 1 x. This is therefore the change in weekly salary for every increase in 1 year of work experience. Thus, the answer is C.

                    @Teach4Success.com

Chapter 6: Nailing Graphs

<u>**KNOW THE SLOPE**</u>:

Slope is rise in a graph over the run in a graph.

Slope formula: (y2 - y1) / (x2 - x1)

Example: Find the equation of the line that passes through (0,5) and (5,0).

Remember: use the basic outline y = mx + b

Thus, using our slope formula:

M = (y2 - y1) / (x2 - x1) = (0 - 5) / (5-0) = -5/5 = -1

Note that where x is 0, this is the y-intercept; in other words, wherever it passes through the origin, that is the y-intercept or your "b" value.

Therefore, the equation of this line is **y = -x + 5**

<u>**Parallel Lines**</u>:

Parallel lines have the same slope or "m" value. Thus an example of a line parallel to the line above is y = -x + 3, as they both have the same negative slope of -1.

<u>**Perpendicular lines**</u>:

Perpendicular lines have negative reciprocal slopes. For instance, if the slope of one line is m, the slope of the other would be -1/m.

<u>**Shifts**</u>:

Note that for shifts,

y = 5(x-3) + 5. This graph is shifted to the right 3 and up 5.

y = 5(x+3) -5. This graph is shifted to the left 3 and down 5.

<u>**Vertex**</u>:

Whenever you are given 3(x-3)(x-5), the vertex is always in between the 2 zeros. So, given this, x = 3 and x = 5, so the vertex is at x = 4.

Practice:

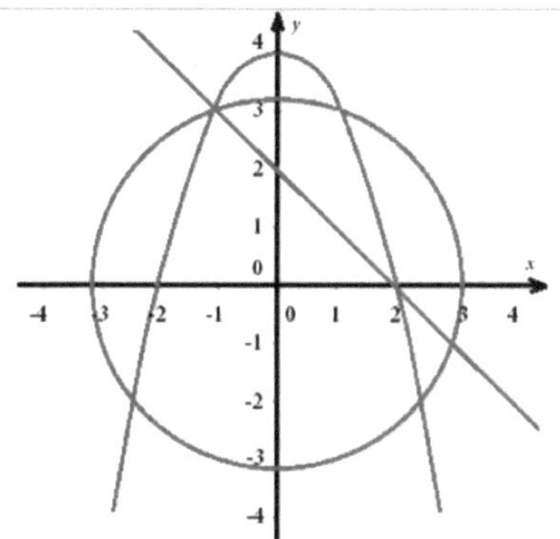

1

A system of three equations and their graphs in the xy-plane are shown in the picture above. How many solutions does the system have?

$$\begin{cases} x^2 + y^2 = 10 \\ y = 4 - x^2 \\ y = 2 - x \end{cases}$$

(A) One

(B) Two

(C) Three

(D) Four

To solve this, look for the point at which all 3 parts of the graph intersect. Doing this gets us the only point at (-1, 3). Thus, the answer is A.

2

What is the vertex of parabola $y = (2x + 6)(x + 1)$?

(A) (3, 1)

(B) (−3, −1)

(C) (−2, −2)

(D) (2, −2)

Remember what we learned. Rearranging this gets 2(x+3)(x+1). Thus the roots are x = -3 and x = -1. So, the vertex must be at x = -2. This means that the answer must be C because it is the only answer choice with -2 at the x-value for the vertex.

@Teach4Success.com

CHAPTER 7: NAILING CENTER OF DATA

KNOW THE MEAN:

The mean is the average. And the average is given by the sum of all terms divided by the number of terms.

For instance, given the terms 5, 6, 7, 8, and 9.

Sum: 35
# of Terms: 5

Mean: 35/5 = 7. Thus, the mean is 7.

KNOW THE MEDIAN:

The median is the middle of all data points. You must ensure that the data points are in order before you find your middle term.

For example,

Given the data set: 1, 2, 3, 4, 4, 5.

This is done as follows:

~~1~~, 2, 3, 4, 4, ~~5~~   Strikethrough the first and last terms and work your way in,

~~1~~, ~~2~~, 3, 4, 4, ~~5~~

Now, we are down to two terms. Once you are down to two or one term you have found your median. In this case, since we have 2 terms, we must average them. This gets: (4+3)/2 = 7/2 = 3.5. Thus 3.5 is the median.

FINDING MEDIAN IN COMPLEX SITUATIONS:

If you are given a set of a ton of numbers, and asked to find the median, it is done as follows.

Say, there are 3001 people waiting in line. The median would be the 1501th term. Why? Because since there are an odd number or even number of terms, the formula for the median is **TERMS/2 + 0.5**. So:

3001/2 + 0.5 = 1500.5 + 0.5 = 1501. Therefore, the median is the 1501th person waiting in line.

<u>KNOW THE MODE:</u>

The mode is just the number that repeats the most. So, in

1, 2, 3, 4, 5, 5. The mode would be 5 because it repeats twice and all other numbers only appear once.

<u>Range:</u>

The range of the data set is just the maximum value minus the minimum value. So, that would be in the above example, 5-1 = 4.

<u>STANDARD DEVIATION:</u>

Standard deviation is how much the values differ from the mean. The more they differ, the higher the standard deviation.

For example,

SET A: 1, 2, 3, 4, 4, 5
SET B: 1, 2, 3, 98 99, 100.

SET B has a higher standard deviation. Why? Because the data values are more spread out. The first set has a mean of 3.8, while the second set has a mean of 60.6. All the data values in the second set are very spread out from the mean.

<u>MARGIN OF ERROR:</u>

The Margin of error is how far off the sample mean is from the true mean of the population. Generally, the margin of error increases while the sample size decreases. But, as the standard deviation increases, the margin of error increases as well.

1

A farmer randomly selected 40 ducks on the farm and measured their weight. The mean weight in the sample was 24 pounds, and the margin of error was 2.8 pounds. The farmer intends to replicate the survey and will attempt to get a smaller margin of error. Which of the following samples will most likely result in a smaller margin of error for the estimated mean weight of ducks?

(A) 25 randomly selected ducks on the farm

(B) 25 randomly selected animals on the farm

(C) 100 randomly selected ducks on the farm

(D) 100 randomly selected animals on the farm

```
The best answer is C. To get a smaller margin of error for the MEAN WEIGHT OF
DUCKS, we need to increase the sample size OF THE DUCKS.
```

2

In the class, the mean height of boys is 173 cm, and the mean height of girls is 165 cm. Which of the following must be true about the mean height, m, of all students in the class?

(A) $m < 169$

(B) $m = 169$

(C) $m > 169$

(D) $165 < m < 173$

```
Whenever you add means, the new mean MUST BE in between the original means.
Therefore, the best answer is D.
```

                              @Teach4Success.com

Chapter 8: NAIL RATES AND UNIT CONVERSIONS

The key to this is to get your ratios to cancel out by having one unit on top and then canceling it with a conversion on the bottom. For example, how many seconds are in an hour?

1 hour * (60 min/hour) * (60 seconds/min) = 1 * 60 * 60 = 3600 seconds.

In this example, all we did is find the respective conversion by canceling out the numerator with its respective denominator.

1

John rides a bike at the average speed of 4 meters per second. He rides a maximum 12 hours every day. How many days does he need to pass 920 miles if he rides the max time everyday? (Note: 1 mile = 1.6 kilometers)

(A) 6

(B) 7

(C) 8

(D) 9

The best answer to this question is D. In order to solve this problem, we must use ratios:

(4 m/s) * (3600 sec/hr) * (12 hr/day) * (1km/1000m) * (1 mile/1.6 km) =

108 miles/day. 920 miles / 108 = 8.5 days. Thus, to cross this, it takes 9 days.

Notice I said 12 hours in a day. This is because the problem explicitly stated that John rides 12 hours maximum in a day.

2

Sally is about to leave to work in 15 minutes. But first she needs to download 10 files of 62 Megabytes each onto her memory stick. Her download internet connection is 3.5 Mbps (Megabits per second). If 1 Megabit equals 0.125 Megabytes, what is the maximum number of files that Sally can fully download within the 15 minutes period?

(A) 5

(B) 6

(C) 10

(D) 0

```
This is a complex problem, but the best answer is B. First set up a unit
conversion:

(3.5 mb/ sec) * (0.125 meg/1mb) * (60 sec/1min) * (15 min)

= 393.75 megabytes in 15 minutes. In each file, there are 62 megabytes.
393.75/62 = 6.35. So, a maximum of 6 files can be made.
```


Chapter 9: Nailing Ratios and Percentages

PERCENT INCREASE/DECREASE:

Formula: (New - old)/old * 100 %

Example: If there are 15 computers in stock on November 1st and 12 computers in stock on November 3rd, by what percent did the stock decrease?

(12 - 15)/15 * 100% = -20% = 20% decrease.

INCLUDED PERCENTAGES:

Convert the percentage to decimal and add or subtract FROM 1 dependent on the problem.

EXAMPLE: If $10 represents the price that includes a 8% fee when you make a purchase at Computer City, what is the price without the fee?

1.08x = 10

X = $9.26

LINEAR VS EXPONENTIAL GROWTH:

If the difference between quantities is constant over time, then you have a linear relationship. If the ratio between quantities is constant over time, then you have an exponential relationship.

Linear: Add 10	Exponential (* 10)
10	10
20	100
30	1000
40	10000

Exponential Growth vs Decay:

A = P(1 + r)^t   Growth
A = P (1-r) ^t   Decay

PROPORTIONS:

Example: There are 12 dogs and 8 cats. If 5 more cats are added to the group, how many more dogs must be added so that 4/5 of the total number of animals in the study are dogs?

20 animals total. 13 cats now. X dogs added to 12.

Solution: $(12 + x)/(20+x) = \%$.

Cross multiplying gives,

$5(12+x) = 80$

$60 + 5x = 80 + 4x$

x = 20

<u>EXAMPLE</u>:

Mia needed exactly 2,950 yen to buy a shirt in Tokyo, so she went to the bank to change dollars into yen. The bank charges a 4% fee on every currency exchange transaction, and Mia gave the bank $26. If the number of yen Mia received was exactly the number she needed to buy the shirt, what foreign exchange rate, in yen per one U.S. dollar, did the bank use for Mia's transaction?

MULTIPLYING 0.04 * 2950 = 118.

CHAPTER 10: NAILING LINES AND ANGLES

PARALLEL VS PERPENDICULAR LINES:

Parallel lines never intersect and have the same slope. Perpendicular Lines form a 90 degree angle at intersection, and the slopes are opposite reciprocals of one another.

ANGLES:
Complementary: Form a 90 degree angle.

Supplementary: Form a 180 degree angle.

Vertical Angles: Equal to one another

LINES CUT THROUGH A TRANSVERSAL:

- CORRESPONDING ANGLES ARE CONGRUENT
- ALTERNATE INTERIOR ANGLES ARE CONGRUENT
- ALTERNATE EXTERIOR ANGLES ARE CONGRUENT

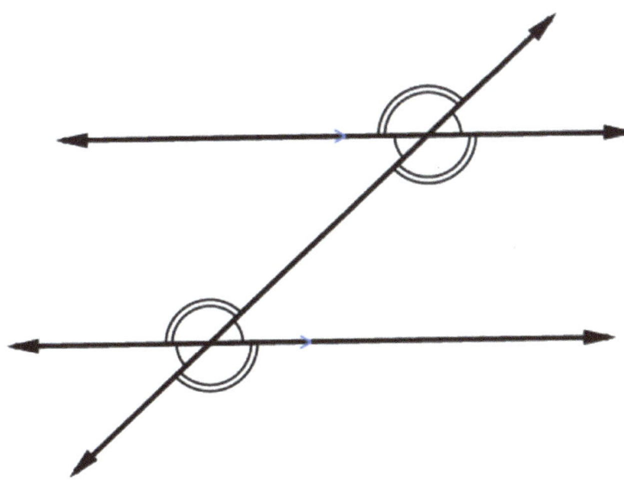

PRACTICE:

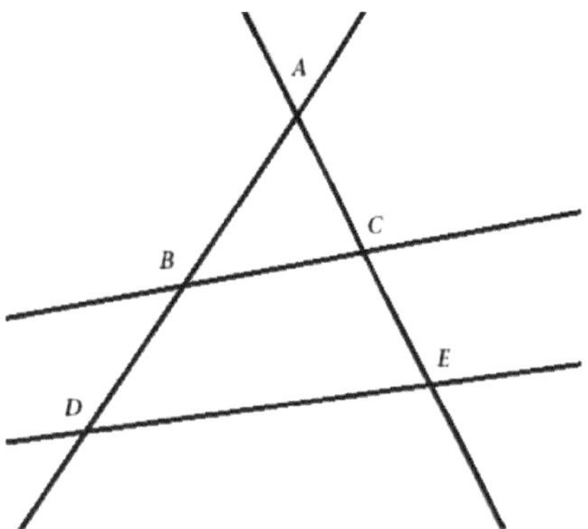

Note: Figure Not Drawn to Scale

1

In the figure above ΔABC is similar to ΔADE. Which of the following must be true?

(A) $BC \perp DE$

(B) $BC \square DE$

(C) $AD \perp AE$

(D) $AD \square AE$

(NOTE THAT THE BOXES MEAN PARALLEL)

ANSWER: B. Since A is a common angle, and we know B is congruent to D as they are corresponding angles, the only correct answer is B because BC and DE must be parallel for the rest of the information to be true.

CHAPTER 11: NAILING TRIANGLES

Area: ½ * b * h

The sum of the interior angles of a triangle is 180.

- EQUILATERAL TRIANGLES:
    - ALL ANGLES ARE CONGRUENT AND EQUAL TO 60. ALL SIDES ARE THE SAME LENGTH.
- ISOSCELES TRIANGLES:
    - TWO OF THE SIDES ARE THE SAME AND TWO OF THE ANGLES ARE THE SAME.
- SIMILAR TRIANGLES:
    - All the angles are the same degree and all the sides are PROPORTIONAL.
- RIGHT TRIANGLES:
    - There is a 90 degree angle, and thus the other two angles are complementary.
    - Pythagorean theorem: the square of the bases added together equals the square of the hypotenuse. (a^2 + b^2 = c^2)

30 60 90 TRIANGLES:

You can split equilateral triangles apart to make these.
The hypotenuse is double the short side and the long side is $\sqrt{3}$ the short side.

45 45 90 TRIANGLES:

You can split squares to make these. (base: 1x base: 1x hypotenuse: $\sqrt{2}$ x)

TRIANGLE INEQUALITY THEOREM:
The third side of a triangle must be greater than the difference and less than the sum of the other two sides

Ex. Two sides of a triangle have the lengths 3 and 4. Which of the following could not be the perimeter of the triangle? (A) 9 (B) 11 (C) 13 (D) 15

The correct answer is D because to have a perimeter of 15, the third side must be 8. However, 3 + 4 is 7, which is less than 8, so they don't make a triangle if we have that perimeter.

                                    @Teach4Success.com

<u>EXAMPLE :</u>

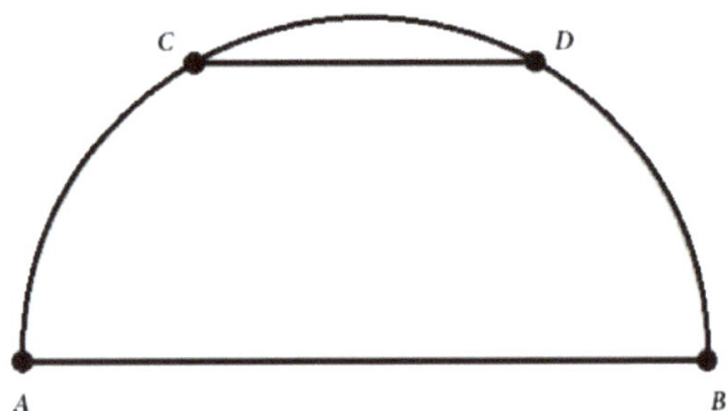

1

The semicircle above has a radius of *r* inches. Chord *CD* is parallel to the diameter *AB*.

If the length of *CD* is equal to the radius of the semicircle, what is the distance between

the chord and the diameter in terms of *r* ?

(A) $\dfrac{r\sqrt{3}}{2}$

(B) $\dfrac{r\sqrt{2}}{2}$

(C) *r*

(D) $\dfrac{2}{3}r$

The correct answer is A. Knowing that CD is equal to the radius of the circle means that you can draw a diagonal line from C to the middle of AB and from D to the middle of AB. This gets you a triangle with CD as the base and two diagonals r. It should look like below.

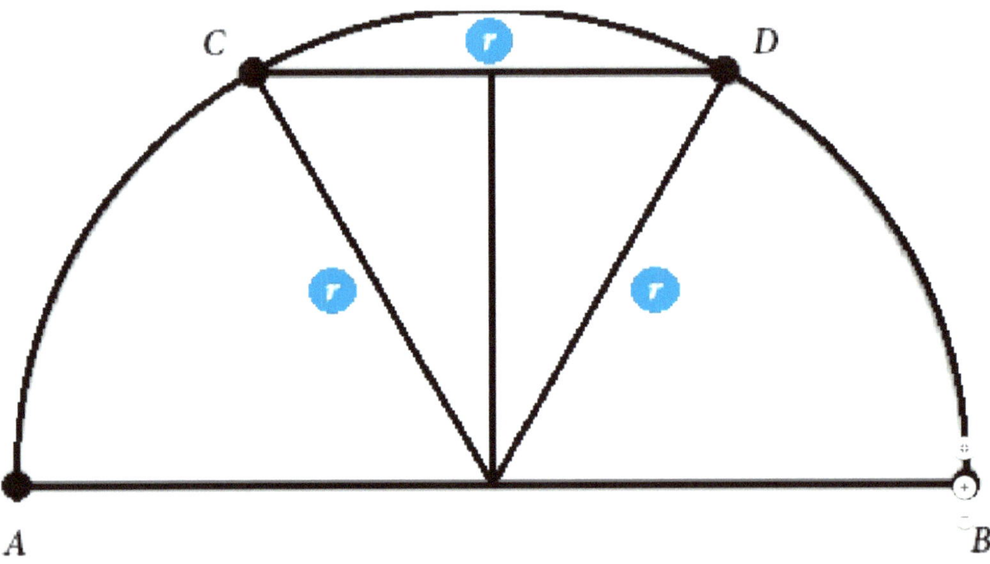

                    @Teach4Success.com

Knowing that all 3 sides are equal means the base angles are both 60. Thus, this is a 60−30−90 triangle. In this case, we know the base is ½ the hypotenuse and the height is sqrt 3 times the base. Since the hypotenuse is r, the base must be 1/2r, thus meaning that the height is √3 * 1/2r = **(r√3)/2** .

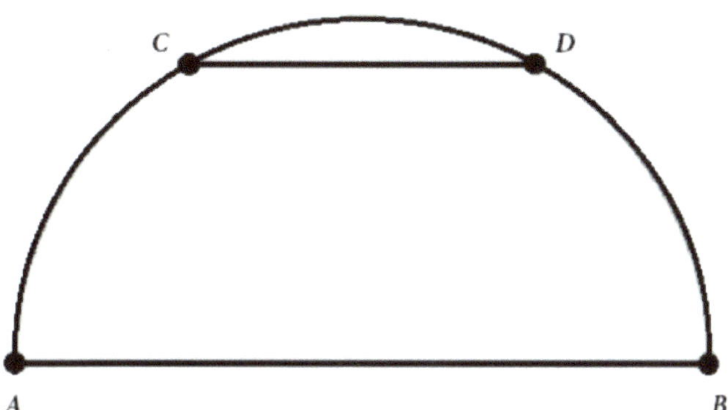

2

The semicircle above has a radius of r inches. The distance between chord CD and diameter AB is $\dfrac{1}{3}$ of the length of AB. What is the area of the trapezoid $ABDC$ in terms of r ?

(A) $\dfrac{2}{3}r^2\left(1+\dfrac{\sqrt{8}}{3}\right)$

(B) $\dfrac{2}{3}r^2\left(1+\dfrac{\sqrt{5}}{3}\right)$

(C) $\dfrac{\pi r^2}{2}-\dfrac{\pi r^2}{2}$

(D) $\dfrac{2}{3}\pi r^2$

The correct answer is B. Drawing in lines should get you the following:

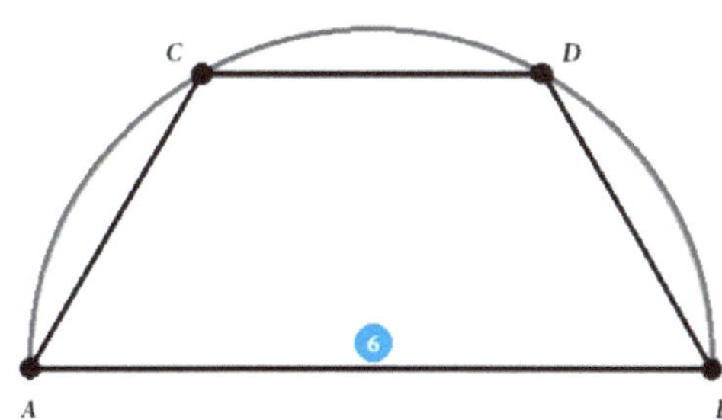

From here, we know AB is 2r. We know the diagonals of the triangle are r. And the height of the triangle is AB/3 = 2r/3. By pythagorean theorem, the length of CD is 2r√5/3. Remember that the area of a trapezoid is ½ h(b1+b2). Plugging 2r in for b2, 2r√5/3 for b1, 2r/3 for h, and dividing by two we would get B.

CHAPTER 12: NAILING CIRCLES

Firstly, it is essential that you're familiar with the following vocabulary:

Radius: The distance from the center to any point on the circumference of the circle

Diameter: The distance from one point on the circle to another point on the circle, but you must go through the center of the circle AND double the radius.

Area of a circle: πr^2

Circumference of a circle: $2\pi r$

Degree: A circle is made up of 360°

Radian: degrees/180 * π. A circle has 2π radians.

Arc Length formula: (Angle/360) * $2\pi r$

Area of a sector formula: (Angle/360) * πr^2

Example:

What is the arc length and sector area of CD if the central angle is 100° in a circle with a radius of 5?

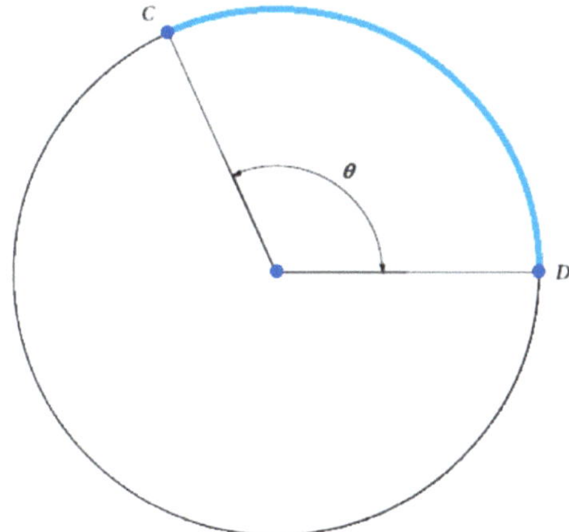

Arc Length: (Angle/360) * $2\pi r$ = 100/360 * $2\pi(5)$ = $50\pi/18$ = $25\pi/9$
Sector Area: (Angle/360) * πr^2 = (100/360) * $\pi(5)^2$ = $125\pi/18$

Inscribed Angles:

Inscribed angles are equal to ½ the measure of its intercepted arc. Inscribed angles that intercept the same arc are equal to each other.

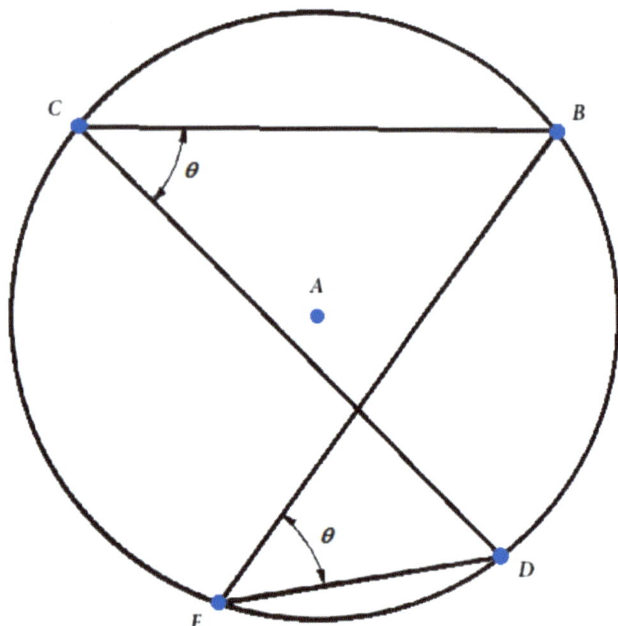

In this example, angle BCD and angle BED are congruent because they are both intercepted by arc BD. If the angle of arc BD is equal to 110°, then angle BCD is equal to 55°.

Finally, know the equation of a circle:

(x-h)^2 + (y-k)^2 = r^2

Practice:

1

The equation of a circle in the xy-plane is shown below. What is the center of a circle?

$$x^2 + y^2 - 2x + 4y = 16$$

(A) $(-1,-2)$
(B) $(1,-2)$
(C) $(-1,2)$
(D) $(1,2)$

Completing the square gives (x-1)^2 + (y+2)^2 = 16 + 1 + 4

(x-1)^2 + (y+2)^2 = 21. Thus, the center is (1, -2). **So B is correct.**

2

The equation of a circle in the xy-plane is shown below. What is the distance between the center of the circle and the point $(1,1)$?

$$x^2 + y^2 + 8x - 4y = 124$$

<u>Completing the square gives:</u>

```
(x+4)^2 + (y-2)^2 = 124 + 16 + 4

(x+4)^2 + (y-2)^2 = 144
```

Thus, the center of the circle is $(-4, 2)$. The distance between this and $(1,1)$ is: $\sqrt{(x1-x2)^2 + (y1-y2)^2}$. Thus, $\sqrt{(-4-1)^2 + (2-1)^2} = \sqrt{26} = 5.1$

CHAPTER 13: NAILING COMPLEX NUMBERS

NOTE:

```
i = √−1
i^2 = -1
```

Practice: Evaluate (7-i)(4+3i)

This gives 28 + 21i -4i -3i^2 = 28 + 17i + 3= **31 + 17i**

PRACTICE OFFICIAL PROBLEMS:

Which of the following is equal to the following expression? (Note: $i = \sqrt{-1}$)

$$\frac{1+i}{1-i}$$

(A) −1

(B) 1

(C) $-i$

(D) i

To get rid of i from the denominator, we need to multiply both the top and bottom by a fancy 1 or (1+i)/(1+i), giving us (1+i)^2/(1+i)(1-i) =

1 + 2i + i^2 / (1 - i^2) = 2i/2 = i. Thus, the answer is **D.**

CHAPTER 14: NAILING TRIGONOMETRY

Know all of these facts:

SINE: opposite/hypotenuse

COSINE: Adjacent/hypotenuse

Tangent: opposite/adjacent

Sinx = cos(□/2-x)

Cosx = sin(□/2-x)

ON THE UNIT CIRCLE, COORDINATE PAIRS ARE WRITTEN AS (Cosx, sinx)

Tanx = sinx/cosx. When the SAT asks you for the tangent of an angle that you know the sine and cosine values for, then you simply put sine over cosine.

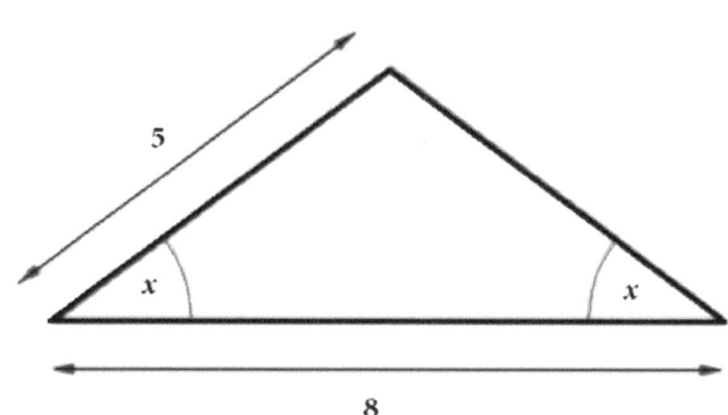

What is the value of sin(x) in the triangle above?

REMEMBER THAT THE PERFECT SCORERS ON THE SAT ALWAYS DRAW IN LINES. Drawing a line in the middle of the triangle, we get a ninety degree angle that makes the bottom portion 4 and the hypotenuse 5. Thus, the height is 3. Sine is opposite over adjacent so ⅗.

2

Given that $\cos(x) = -a$, where x is a radian measure of an angle and $\dfrac{\pi}{2} < x < \dfrac{3\pi}{2}$. If $\cos(w) = a$, which of the following could be the value of w ?

(A) $x + \pi$

(B) $x - \dfrac{\pi}{2}$

(C) $x + 2\pi$

(D) $x - 2\pi$

The correct answer is A. Adding or subtracting 2pi would get you the same value not a different value, so both C and D are out.

Next let us try plugging in pi for x. Cos pi = -1. The question said that cos(x) = -a, so -a = -1, so a =1.

Thus, cos(w) = 1.

We need to find w. cos(w) =1 when w = 2pi or 0. To get this value, we need to add or subtract pi to x. Therefore, the answer is A.

